Read All About
Earthly Oddities

OCEAN CURRENTS

Patricia Armentrout

The Rourke Press, Inc.
Vero Beach, Florida 32964

PHOTO CREDITS
© Dennis Carmichael: pg. 22; © Dick Dietrich: pg. 6; © Bob Firth/Int'l Stock: pgs. 15, 19; © NASA: pgs. 10, 21; © NOAA: pgs. 12, 16; © Jonathan E. Pitt/Int'l Stock: pg. 18; © Marty Snyderman: pg. 9; © Oscar C. Williams: pgs. 4, 7; © Woods Hole Oceanographic Institution: Cover, pg. 13

ACKNOWLEDGMENTS
The author wishes to acknowledge David Armentrout for his contribution in writing this book.

Library of Congress Cataloging-in-Publication Data

Armentrout, Patricia, 1960-
Ocean Currents / by Patricia Armentrout.
p. cm. — (Earthly Oddities)
Includes index.
Summary: Examines the currents of the oceans, how they are mapped, what keeps them moving and how they can affect the weather.
ISBN 1-57103-153-7
1. Gulf stream—Juvenile literature. [1. Gulf stream.]
I. Title II. Series: Armentrout, Patricia, 1960- Earthly Oddities.
GC296.G9A76 1996
551.47'11—dc20 96–2893
CIP
AC

Printed in the USA

TABLE OF CONTENTS

RIVERS IN THE OCEANS

Have you ever taken a canoe trip down a stretch of river? The flow of the river moves the canoe in one direction—down stream. Water is moving in thousands of rivers and streams all over the world. But not all rivers flow on land.

Some say there are rivers in the oceans. These rivers are called **currents** (KER entz). Ocean currents flow in all directions—north, south, east, and west. Some currents even move water from the surface towards the ocean floor and back again.

The gentle current of this river helps the canoeists down stream.

WHAT KEEPS CURRENTS MOVING?

Two factors keep the currents moving: the heat from the sun and the Earth's wind system.

The sun's heat, or energy, warms the ocean water in the **tropics** (TRAHP iks). Warm water is lighter than cold water and flows close to the surface. Cooler water temperatures sink towards the bottom.

The sun's heat warms the ocean water.

Wind-blown waves crash onto shore.

Wind causes the surface water to circulate. Together, the wind and differences in water temperatures cause the ocean water to be in constant motion.

THE GULF STREAM

The Gulf Stream is a warm ocean current in the Atlantic Ocean. The current begins off the southern coast of Florida. The Gulf Stream is fed by the fast-moving waters of the Florida Current.

The Gulf Stream is home to many types of animals that take advantage of its flow. Sea creatures such as sharks, whales, and many fish use the current to cruise the ocean in search of food.

The Gulf Stream is just one current that combines with other currents to create the Gulf Stream system.

Humpback whales travel to the warm waters of the tropics during the winter months.

THE GULF STREAM SYSTEM

The Gulf Stream system is the constant circular flow of several ocean currents. The general course of the system does not change. The boundaries can vary, though, because many currents are always entering and leaving the system.

After collecting warm waters from the Florida Current, the Gulf Stream heads north along the east coast of the United States. At North Carolina, the system of currents turn east across the North Atlantic. Off the coast of Europe the system turns south. It then flows back across the Atlantic Ocean through the **Caribbean** (kar uh BEE un), ending and beginning its cycle again at the Florida Straits.

The Florida Current speeds through the Straits of Florida to join the Gulf Stream.

MAPPING THE GULF STREAM

In 1768 a now-famous American named Benjamin Franklin was the Deputy Postmaster General for the American Colonies. Franklin was often told that it took weeks longer to receive mail shipped from England than it did to ship mail there.

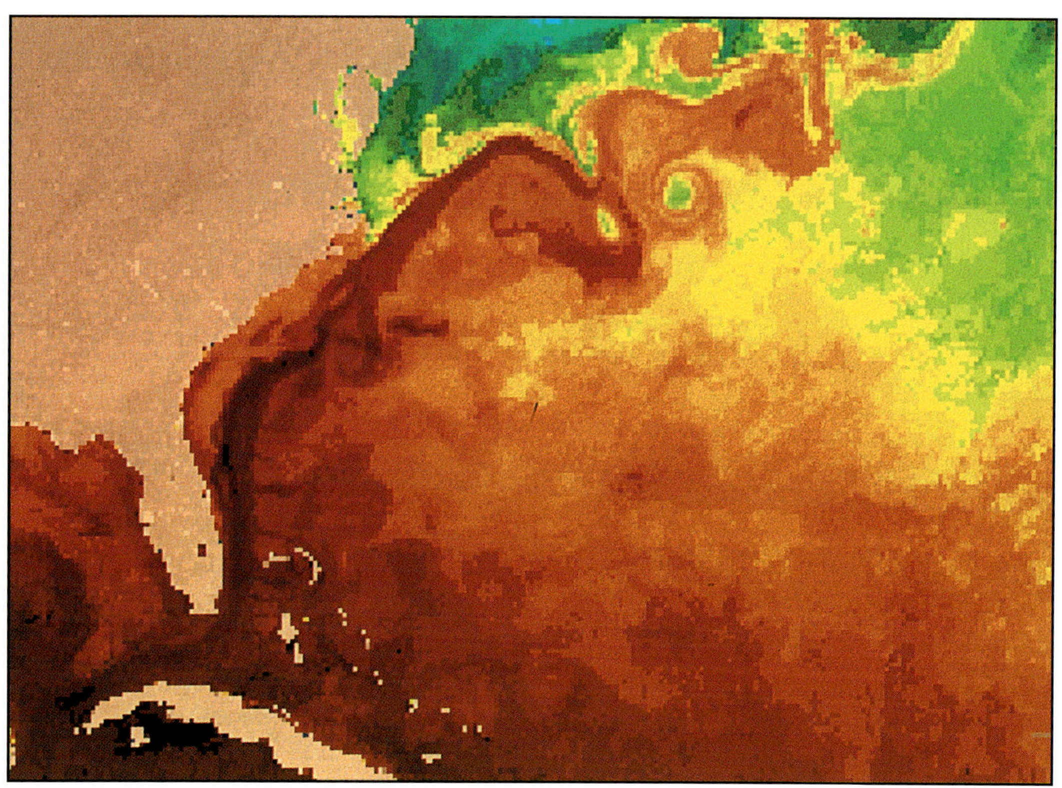

The deep red color in this satellite image shows the warm water flow of the Gulf Stream system.

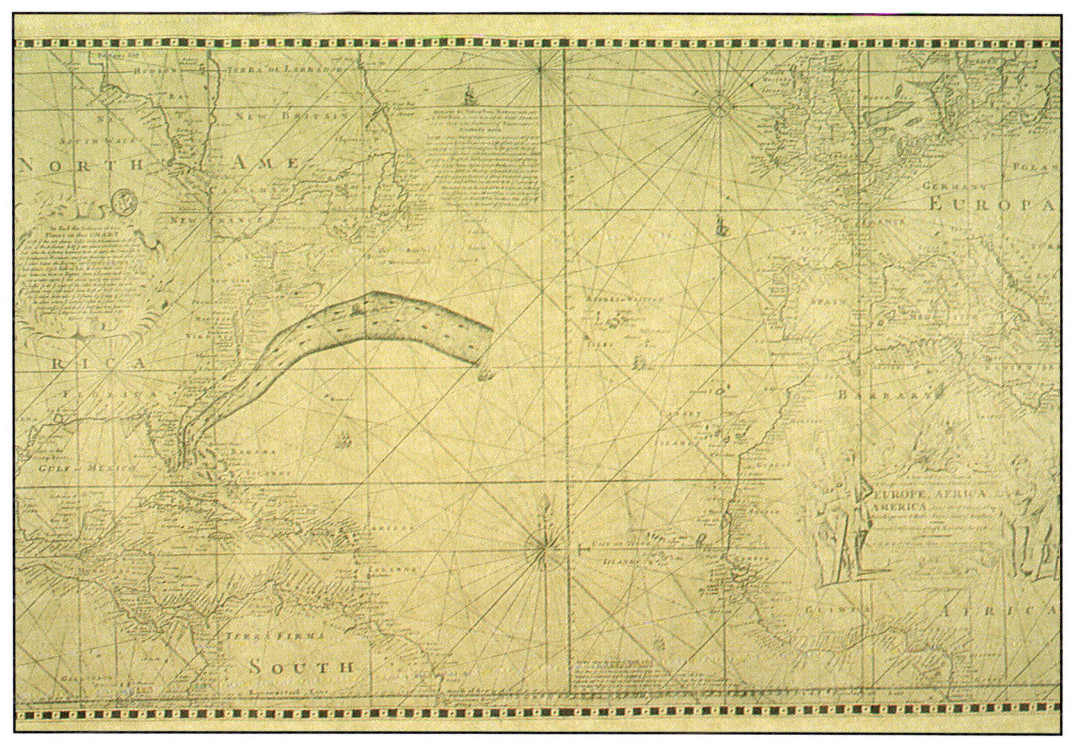

This map shows the original charting of the Gulf Stream.

Franklin researched the problem with ship captain Timothy Folger. Folger told Franklin about the strong currents that flow up the American coastline and east across the Atlantic. The results of Franklin's research led to the mapping of the Gulf Stream in 1769. Eventually captains learned to avoid sailing against the Gulf Stream current on their voyages west.

EL NINO

El Nino (el NEEN yo) is a warm current in the Pacific Ocean. El Nino is a seasonal current. It appears at the end of December and disappears by the end of March. Sailors named the current El Nino, which means "Christ child," because it first appears close to the Christian holiday called Christmas.

El Nino flows south along South America near Peru. Water temperatures rise several degrees above average. Every two to seven years the current drifts farther south.

This small change in water temperature often has a major effect on weather all over the world. The warmer ocean water has caused heavy rainfall and flooding in some areas, and drought and fires in others.

This farm is caught in floodwater after weeks of heavy rain.

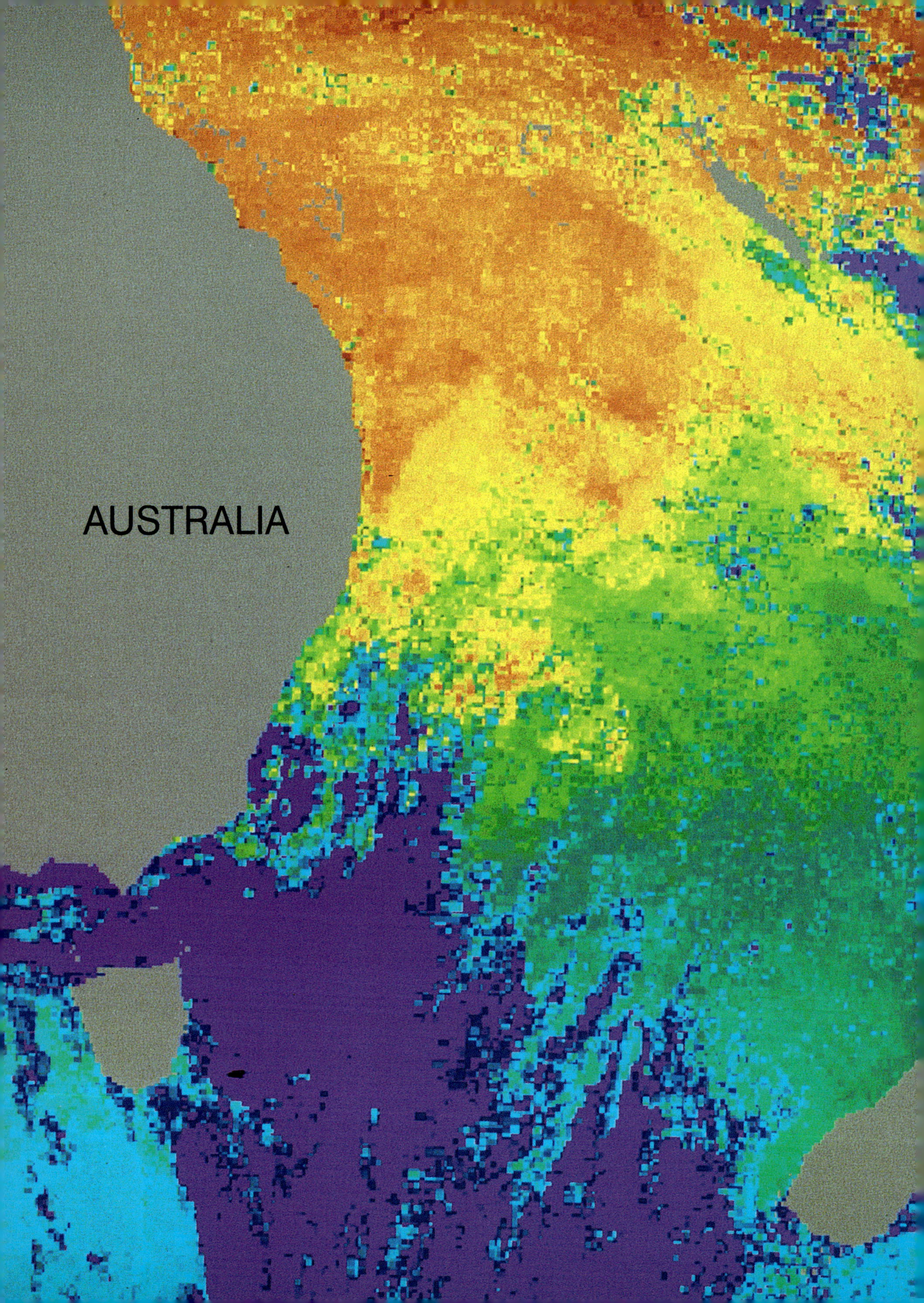
AUSTRALIA

OTHER OCEAN CURRENTS

Many currents flow through our oceans. Some currents have warm water temperatures while other currents are cold.

The East Australian Current brings warm water to Australia's eastern coast as it flows south. Plants and animals in the Great Barrier Reef thrive in the warm ocean water.

The Antarctic Current stays in a circular flow around Antarctica; the waters there are always cold. The Peru Current flows north from the Antarctic area. It brings cool water temperatures and fog to Chile and Peru.

The orange color in this satellite image shows the warm water of the East Australian Current.

CURRENTS AND WEATHER

Weather all over the world is affected by ocean currents. In 1982-1983 the seasonal change in El Nino caused weather disasters. Australia and Africa suffered droughts, dust storms, and fires. Peru had record-breaking rainfall that caused flooding and landslides.

Long periods without rain can cause the earth to dry and crack.

Sand bags are piled high to keep floodwater under control.

The warm waters of the Gulf Stream cause sections of Europe to have a warmer climate than it would have without them. In fact, tropical palm trees grow in some areas of Ireland and Scotland, which would not be possible without the warming effect of the Gulf Stream.

ONE WORLD OCEAN

What do you think of when you hear the word "earth?" Do you think of soil and rock? That is how the word is defined, but the Earth's surface is mostly water. Water covers over 70 percent of our planet.

Ocean currents help sea animals to **migrate** (MY grayt) in search of food. Currents help ship captains **navigate** (NAV eh gayt) the open sea. Ocean currents mix water temperatures and move water to every part of the globe, which is why we call our oceans "One World Ocean."

All ocean waters mix from the flow of currents and become "One World Ocean."

GLOSSARY

Caribbean (kar uh BEE un) — a sea in the warm western section of the Atlantic Ocean

currents (KER entz) — constant movement of water

El Nino (el NEEN yo) — a warm, seasonal current in the Pacific Ocean that changes flow every two to seven years

migrate (MY grayt) — to move from one place to another mainly in search of food

navigate (NAV eh gayt) — to sail or steer a course on a ship or aircraft

tropics (TRAHP iks) — the area between 23$\frac{1}{2}$ degrees north and south of the equator where the sun shines directly overhead

Grunts swim in schools and are common in the warm waters of the Caribbean.

INDEX